# The
# TEXAS
# FIRE ALARM
# INSPECTOR

Kenneth R. Stephens CET

authorHOUSE®

AuthorHouse™ LLC
1663 Liberty Drive
Bloomington, IN 47403
www.authorhouse.com
Phone: 1-800-839-8640

Published by AuthorHouse  05/02/2014

ISBN: 978-1-4969-0519-2 (sc)
ISBN: 978-1-4969-0520-8 (e)

## AUTHOR'S FOREWORD

This book reflects my understanding of the National Fire Alarm and Signal Code (NFPA 72) 2010 edition and the National Fire Protection Association 2009 Seminar in which I received training in the subject matter of fire alarm inspections. The National Fire Protection Association nor their employees have any liability concerning the accuracy of this book. This book is my understanding and viewpoint only. The National Fire Protection Association owns all rights to NFPA 72 and the NFPA 2009 Seminar.

In the past, I have been asked questions about how and why I do fire alarm inspections the way I do them. This book is my testimony for the record as a professional based upon my training from NFPA as to how and why my fire alarm inspections are performed.

Kenneth R. Stephens
Dallas County, Texas

*Kenneth R. Stephens CET*

NATIONAL FIRE ALARM AND SIGNALING CODE
NFPA 72
2010 EDITION

## INTRODUCTION

The Texas Fire Alarm Inspector reviews Chapter 14 of NFPA 72, The National Fire Alarm and Signaling Code, 2010 edition. For brevity I will refer to this code as NFPA 72, 2010. The Texas Fire Alarm Inspector will be focused on inspections as opposed to testing and maintenance. Inspections in Chapter 14 is a separate category. Testing is a separate category. Maintenance is a separate category.

In previous years of the alarm industry, there was little distinction between the categories. A service technician did it all and combined all of the procedures into one service report.

Today with the introduction of new professional guidelines and legal requirements, the old service report has given way to a list of new specialized reports.

*Kenneth R. Stephens CET*

NFPA 72, 2010
CHAPTER 14

INTRODUCTION
TERMS USED IN THIS BOOK.

Before reading this book, it is important that the reader know and understand the following terms:

| | |
|---|---|
| NFPA | NATIONAL FIRE PROTECTION ASSOCIATION. |
| NFPA 72 | NATIONAL FIRE ALARM and SIGNALING CODE. |
| 2010 | The 2010 edition of the NATIONAL FIRE ALARM CODE. |
| CHAPTER 14 | INSPECTION, TESTING, and MAINTENANCE CHAPTER of the NFPA 72 |
| NFPA 101 | NFPA LIFE SAFETY CODE |
| NFPA 70 | NATIONAL ELECTRICAL CODE (NEC) |
| NFPA 13 | STANDARD for INSTALLATION of SPRINKLER SYSTEMS |
| NFPA 25 | INSPECTION and TESTING of WATER BASED SYSTEMS |

TERMS that the author avoids:
Sprinkler System.
The author uses the term Fire Sprinkler System because of the confusion on the part of the public with lawn sprinklers.

AC
The author uses more specific terms.
        VAC= Voltage Alternating Current.
        HVAC= Heating, Ventilation, Air Conditioning.

*Kenneth R. Stephens CET*

## NFPA 72, 2010 EDITION

## INTRODUCTION

The Inspector

Under NFPA 72, 2010; the inspector does inspections. The tester does tests. The maintenance person does maintenance. In previous years of the alarm industry these jobs were all done by one person at the same time.

Clearly, one technician may be qualified to perform all of these duties but new procedures and requirements ask for specialized reporting that must be done separately.

Inspectors do more frequent inspections than the frequency of tests required for testing. Inspections are "no touch" visual reports. The inspection format is looking for changes to the fire alarm system and damage to the system. Inspections are performed quarterly or semiannually, as well as annually.

*Kenneth R. Stephens CET*

NATIONAL FIRE PROTECTION ASSOCIATION
MEMBERSHIP

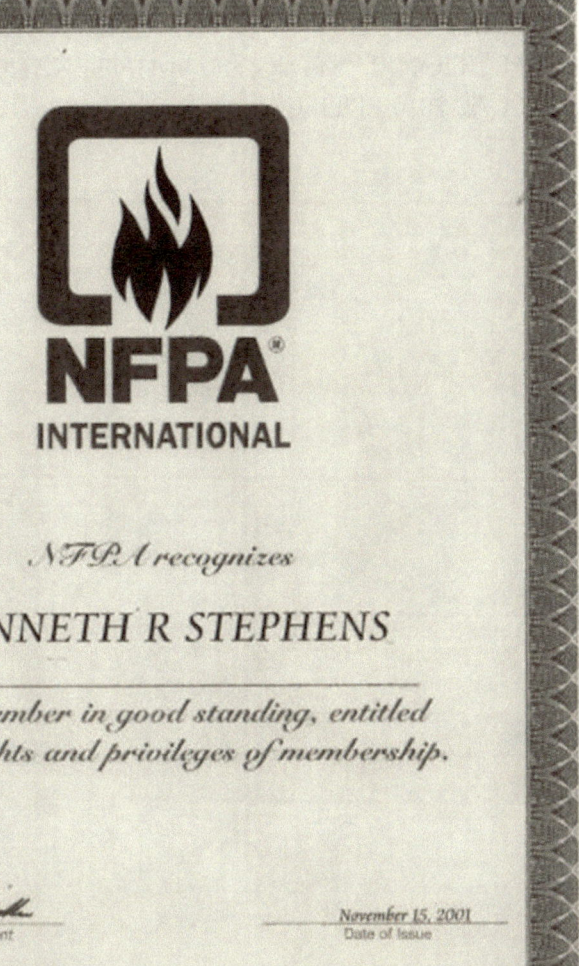

*Kenneth R. Stephens CET*

NFPA 2009 TRAINING COURSE
AMERICAN AIRLINES TRAINING CENTER
FORT WORTH, TEXAS
12/3/2009
7 HOURS
INSPECTION, TESTING, and MAINTENANCE OF FIRE ALARMS
NFPA 72, 2010 EDITION

Inspection, Testing and Maintenance of Fire Alarms

Certificate of Attendance

**KENNETH STEPHENS**
FORT WORTH, TX  12/3/2009
.7 CEUs or 7 hours

President, National Fire Protection Association

NFPA 2009 Seminar

# DEFINITIONS

# NFPA 2009 SEMINAR

NFPA 2009 Seminar

Definitions

Codes are the requirements.

Standards are the way to meet code requirements. Listed and Labeled Equipment are displayed in the directories of Underwriters Laboratories and Factory Mutual.

Listing fulfils the label (very specific).

Requirements use the word "shall" (enforceable).

Annexes use the word "should" (not enforceable).

NFPA 2009 SEMINAR

NFPA 72, 2010 CHAPTER 3

NFPA 2009 Seminar

DEFINITIONS

Approved.     Acceptable to the authority having jurisdiction.

The authority having jurisdiction may base acceptance on compliance with NFPA or other appropriate standard.

The authority having jurisdiction may also refer to the listings or labeling practices of an organization that does product evaluations.

NFPA 72, 2010 Chapter 3

# NFPA 72, 2010 CHAPTER 3

NFPA 2009 Seminar

DEFINITIONS

ALARM= Signal that indicates immediate action is necessary

SUPERVISORY= Normal or Off Normal signal

TROUBLE = Fault Indication

NFPA 72, 2010 Chapter 3

NFPA 72, 2010 CHAPTER 3

NFPA 2009 Seminar

DEFINITIONS

Labeled.     Equipment or materials to which has been attached a label, symbol, or other identifying mark of an organization that is acceptable to the authority having jurisdiction and concerned with product evaluation, that maintains periodic inspection of production of labeled equipment.

Listed.      The means for identifying listed equipment that meets the appropriate standard or has been tested and found suitable for a specific purpose.


NFPA 72, 2010 Chapter 3.

# NFPA 72, 2010 CHAPTER 3

NFPA 2009 Seminar

DEFINITIONS

SHALL.        Indicates a mandatory requirement

SHOULD.     Indicates a recommendation, that which is advised, but
             not required.


NFPA 72, 2010, Chapter 3.

NFPA 2009 SEMINAR
12-3-2009
7 HOURS

NFPA 2009 Seminar

# RECORDS

NFPA 72, 2010, CHAPTER 10
CHAPTER 14

NFPA 2009 Seminar

Inspection, Test, and Maintenance Records.
NFPA 72, 2010, Chapter 10

Retain all records until one year after next test.
Records must be promptly provided to the AHJ upon request.
Supervising station records must be maintained for not less than one year.

NFPA 72, 2010, Chapter 14

ITM records retained until next test plus 1 year. Paper or electronic media permitted.
ITM records must include information required by Section 6.2.3 of Chapter 14.
Simulation notes will be available.
Special requirements for supervising stations.

NFPA 72, 2010 Chapter 10 Section 18.1.2
APPROVAL DOCUMENTATION (REQUIRED)

Type of service
Specifications
Drawings
Operations matrix
Battery calculations
NAC (Notification Appliance) voltage drop calculations
Record of Completion

NFPA 72, 2010   CHAPTER 10
               CHAPTER 26

NFPA 2009 Seminar

Final documentation kept at FACP.
NFPA 72, 2010 Chapters 10 and 26.

Record of Completion
Record Drawings (As-built)
Written sequence of operations
Operations and Maintenance Manual
Statement of Compliant Installation
Record of Inspection, Test, and Maintenance

If documents are stored in a separate cabinet that cabinet must be labeled: "FIRE ALARM DOCUMENTS".

NFPA 2009 SEMINAR
NFPA 72, 2010 EDITION

NFPA 2009 Seminar

# INSPECTIONS

NFPA 72, 2010 CHAPTER 14

*Kenneth R. Stephens CET*

NFPA 2009 Seminar

INSPECTIONS

CONTROL EQUIPMENT
  Fuses
  Interfaced Equipment
  Lamps and LED's
  Primary Power Supply


MONITORED—ANNUAL INSPECTION
UNMONITORED—WEEKLY INSPECTION


NFPA 72, 2010, Chapter 14 Section 3.1, item 1
NFPA 72, 2010, Chapter 14 Section 3.1 item 2
Tables

NFPA 72, 2010   CHAPTER 14
             CHAPTER 10

NFPA 2009 Seminar

INSPECTIONS

BATTERIES

Lead Acid
DRY CELL—Primary             MONTHLY

NICKEL-CADMIUM (NICAD)
SEALED LEAD-ACID SEMIANNUALLY

Batteries must have manufacturer's marked date of manufacture (month and year).

NFPA 72, 2010, Chapter 14 Section 3.1, item 3
                Chapter 10 Section 5.9.1.1

NFPA 72, 2010 CHAPTER 14

NFPA 2009 Seminar

INSPECTIONS

MANUAL PULL FIRE ALARM BOXES. SEMIANNUALLY.
NFPA 72, 2010, Chapter 14 Section 3.1 item 9.e

SYSTEM SMOKE DETECTORS SEMIANNUALLY


Spot Smoke Detectors
Smoke Beams

NFPA 72, 2010 Chapter 14 Section 3.1 item 9.h


AIR SAMPLING SMOKE DETECTORS SEMIANNUALLY

NFPA 72, 2010, Chapter 14, Section 3.1, item 9.a

HEAT DETECTORS            SEMIANNUALLY

NFPA 72, 2010, Chapter 14, Section 3.1 item 9.f

NFPA 72, 2010 CHAPTER 14

NFPA 2009 Seminar
INSPECTIONS
OPTICAL FIBER CONNECTIONS. INSPECT ANNUALLY.

METALLIC CONDUCTORS.     NO VISUAL INSPECTION
                        REQUIREMENTS in NFPA 72.

    NOTE: T-Taps are not allowed.
    EXCEPTION: T-TAPS are allowed on Class B SLC loops.

NFPA 72, 2010, Chapter 14 Section 3.1 item 6

NFPA 72, 2010 CHAPTER 14

*Kenneth R. Stephens CET*

NFPA 2009 Seminar

INSPECTIONS

FIRE SPRINKLER DEVICES

WATERFLOW ALARMS                    QUARTERLY

NFPA 72, 2010, Chapter 14, Section 3.1, item 9.j

VALVE SUPERVISORY DEVICES QUARTERLY

NFPA 72, 2010, Chapter 14, Section 3.1, item 9.i

OTHER SUPERVISORY DEVICES QUARTERLY

NFPA 72, 2010, Chapter 14, Section 3.1, item 9.i

NFPA 72, 2010 CHAPTER 14

*Kenneth R. Stephens CET*

NFPA 2009 Seminar

INSPECTIONS

SUPPRESSION SYSTEM SWITCHES SEMIANNUALLY

NFPA 72, 2010, Chapter 14, Section 3.1 item 9.d

NFPA 72, 2010, CHAPTER 14

*Kenneth R. Stephens CET*

NFPA 2009 Seminar

INSPECTIONS

RADIANT ENERGY SENSING DETECTORS      QUARTERLY

NFPA 72, 2010, Chapter 14, Section 3.1 item 9.g

NFPA 72, 2010 CHAPTER 14

*Kenneth R. Stephens CET*

NFPA 2009 Seminar

INSPECTIONS

DUCT SMOKE DETECTORS        SEMIANNUALLY

NFPA 72, 2010 Chapter 14, Section 3.1, item 9.b

NFPA 72, 2010 CHAPTER 14

NFPA 2009 Seminar

INSPECTIONS

NOTIFICATION APPLIANCES      SEMIANNUALLY

NFPA 72, 2010, Chapter 14 Section 3.1, item 12

NFPA 72, 2010 CHAPTER 14

*Kenneth R. Stephens CET*

NFPA 2009 Seminar

INSPECTIONS

ANNUNCIATORS     SEMIANNUALLY

NFPA 72, 2010, Chapter 14, Section 3.1, item 8.

# NFPA 2009 SEMINAR

NFPA 2009 Seminar

INSPECTIONS, TESTS, and MAINTENANCE

# NFPA 2009 SEMINAR—NOTES

NFPA 2009 Seminar                     Notes

Why Inspect, Test, and Maintain?

1. Ensure initial installation meets code requirements.
2. Ensure changes to the fire alarm system do not adversely affect performance.
3. Ensure fire alarm system remains operational.
4. Ensure discovery of failed components within prescribed limits.
5. NFPA 72 is not optional.

# NFPA 2009 SEMINAR—NOTES

NFPA 2009 Seminar |Notes

Testing

Physical manipulation of a component or system function to determine operational status.

Maintenance

Action taken to maintain a component in service.

> Follow manufacturer's instructions.
>
> Frequency of maintenance depends on ambient conditions.
>
> Apparatus requiring rewinding or resetting shall be restored immediately after testing.

## NFPA 2009 SEMINAR—NOTES

## INSPECTION of SPECIAL HAZARDS.

Coordinate with releasing company before testing special hazards/ fire suppression equipment.

Verify releasing circuit by simulation.

# NFPA 2009 SEMINAR—NOTES

NFPA 2009 Seminar

Supplementary Equipment.

Is supplementary equipment subject to the requirements of NFPA 72 Chapter 14?
Yes, but only to the extent that it affects the normal operation of the fire alarm.

Supplemental signals are not required by the Code.

Examples of Supplemental Equipment;
Smoke Dampers.
Elevator Recall.

NFPA 72, 2010,   CHAPTER 14
                 CHAPTER 10

NFPA 2009 Seminar

## QUALIFICATIONS

Service personnel shall be qualified

Qualifications may include:

Factory trained and certified for the make and model being serviced.

National certification approved by AHJ.

Registered or licensed by AHJ.
Employed by a listed service company.

Provide evidence to AHJ of qualifications.

NFPA 72, 2010   Section 2.2.5 of Chapter 14
                Section 4.3.1 of Chapter 10

# THE TEXAS FIRE ALARM INSPECTOR
# NICET CERTIFICATIONS

KENNETH R STEPHENS
CERTIFIED ENGINEERING
TECHNICIAN

TEXAS FIRE ALARM INSPECTOR


NICET CERTIFICATIONS

ORIGINAL APPLICATION
WORK ELEMENTS EXAMINATION
PERFORMANCE VERIFICATION
WORK HISTORY


NOTE: SOCIAL SECURITY NUMBERS ARE BLOCKED OUT.

2005
NICET TECHNICIAN APPLICATION FORM
PART 1 APPLICATION

**National Institute For Certification In Engineering Technologies**
sponsored by the National Society of Professional Engineers
www.nicet.org

**NICET TECHNICIAN APPLICATION FORM**

**Part I: Applicant Information**

Section 1- Personal Information (Please Print Clearly or Type)

| A | B |
|---|---|
| ☒ Mr. ☐ Ms. STEPHENS KENNETH R. | Present Employer: ADT SECURITY SERVICES INC |
| Last Name    First Name    Middle Initial | Company Name |
| Home Address: Mailing Address | Business Address: |
| 25 Highland Village No. 100 | 2403 Lacy Lane |
| Street    Apt. | |
| Dallas    Tx.    75205 | Carrolton    Tx. 75006 |
| City    State    Zip Code  +4 | City    State    Zip Code  +4 |
| Social Security Number _____ | Present Position Title: Service Technician/Inspector |
| Home Phone 972-877-0334 | Work Phone 972-246-6900 |
| Area Code    Number | Area Code    Number    Ext. |
| E-mail    krr@ncmx.net | Cell Phone 214-6 |
| May NICET use your e-mail address to contact you?  ☒ yes ☐ no | Area Code    Number |
| To which address should NICET mail all correspondence? | Fax |
| ☒ Home  ☐ Business | Area Code    Number |

C. Has your last name changed since you submitted your last application? ☒ no ☐ yes, former last name _____

D. Are you a student in an Engineering or Engineering Technology degree program? ☒ no ☐ yes (provide more information below)

Institution _____ City/State _____

Name of degree program _____ Expected Graduation Date _____

**Section 2 – Objectives**

A. Check off the one item below that best describes your examination goal:
I am applying for the following certification(s) and intend to submit all documents required for certification.

| Program (field/subfield) | Level or Grade |
|---|---|
| Fire Protection Engineering Technology | |
| Fire Alarm Systems | II |
| Fire Protection Engineering Technology | |
| Inspection and Testing of Water Based Systems | II |

I am applying to take an examination only and will notify NICET if I decide to pursue certification.

B. Please check off the box below that applies to you:
not tested before  ☒ tested before, but not NICET-certified  ☒ am NICET-certified, certification # 114545

**Section 3 - Applicant's Statement of Understanding**
(Your signature and the date must appear after the following statement; otherwise this application will not be accepted.)

I certify that all information given on my application and any supporting materials is correct, factual, and complete. I understand that any misrepresentation of information can result in the rejection of this application or the revocation of any certificate NICET has issued in my name. Further, I certify that I have read and understood the instructions for this application and that I have read, understood, and I accept the conditions set forth in the "Conditions of Application for Technicians".

Signature: *Kenneth R. Stephens*    Date: 9-21-05

**Examination Fees are Non-Refundable**

A check or money order, payable to NICET, must be enclosed with any testing application. Fees are listed on our Website, www.nicet.org, or can be obtained by calling 888-476-4238 or 703-548-1518. Payments may be deductible under applicable provisions of the Internal Revenue Code (i.e., as educational or business expenses); however, payments are not deductible as charitable expenses. The examination fee is non-refundable.

Mail this form with fee payment to: NICET, c/o Bank of America, Dept 9637, Washington, DC 20055

| Employer ID | Postmark Date | Spec. Cont. | ADA | Lockbox #1 | Amt Paid | Lockbox #2 | Amt Paid |
|---|---|---|---|---|---|---|---|

SOCIAL SECURITY NUMBER IS BLOCKED OUT.

**SOCIAL SECURITY NUMBER IS BLOCKED OUT.**

*Kenneth R. Stephens CET*

2005

NICET TECHNICIAN APPLICATION
PART 2. EXAMINATION

Name: Kenneth R. Stephens

Social Security No.
Blocked
Social Security Number: _____

Work Element

## NICET Technician Application, Part II: Examination
(Please print clearly or type)

### Section 1 - Examination Scheduling

A. Applicants for examination at a regular NICET test center must provide the information requested below. (Locations, dates, and postmark deadlines are listed on NICET's Website. If you do not have access to the Internet, please contact NICET at 888-476-4238 or 703-548-1518.)

1. 1st choice: Frisco Tx.  City & State    Tx 11 Test Center Code    Exam Date: 02 Mo. 26 Day 05 Year

2. 2nd choice: Amarillo Tx.  City & State    Tx 1 Test Center Code    Exam Date: 02 Mo. 26 Day 05 Year

(It is not required that you select a 2nd choice; however, if your 1st choice is full or you miss the postmark deadline for the 1st choice and you have no 2nd choice, your test will be scheduled for the next test date at the requested location.)

B. If you are testing at two different locations in the same cycle, please check here: ___ yes, testing twice

C. The date of my last exam was October 16, 2004 . (Note: If you are not yet certified and you have not tested in 5 years, your test history will be deleted. Refer to Policy #26 on NICET's Website at www.nicet.org.)

### Section 2 – Work Element Selection and Verification

List below the work elements that you are requesting for this exam. The field code # is printed on the front cover of the program detail manual and is necessary to fully identify the work element. The work element ID numbers and titles are in the "Work Element Listing" in each manual. The maximum number of elements you can select for one exam is 34.

Please write with care; you are responsible for the correct identification of work elements for your exam.

| 3-Digit Field Code No. | 5-Digit Work Element ID No. | Work Element Title | Verifier's Initials* | 3-Digit Field Code No. | 5-Digit Work Element ID No. | Work Element Title | Verifier's Initials* |
|---|---|---|---|---|---|---|---|
| 003 | 31011 | Basic Metric Units / Conversions | BS | 003 | 41001 | Inspection Procedures | BS |
| 003 | 31003 | Basic Wiring | BS | 003 | 41002 | NFPA Standard Definitions | BS |
| 003 | 31004 | Devices and Components | BS | 003 | 41003 | Basic Components | BS |
| 003 | 33007 | Detector Spacing | BS | 003 | 41006 | Role of the Inspector | BS |
| 003 | 33008 | Power Supplies | BS | 003 | 41007 | On-Site Inspectional Testing Req. | BS |
| 003 | 33020 | Central Station Fire Alarm | BS | 003 | 41008 | General System Terminology | BS |
| 003 | 33021 | Manual Fire Alarm/Guards Tour | BS | 003 | 42002 | Foam Water System | BS |
| 003 | 35001 | Surveys - Fire Detection Systems | BS | 003 | 42009 | Fire Pumps | BS |
| 003 | 35002 | Shop/Riser Drawings | BS | 007 | 37003 | System Reliability | BS |
| 003 | 35003 | As Built Drawings | BS | 003 | 37004 | Hazard Analysis | BS |
| 003 | 35004 | Principles of Smoke Movement | BS | 003 | 37005 | Avoidance of Nuisance Alarms | BS |
| 003 | 35005 | Supplementary Circuits | BS | 007 | 37006 | Special Protection | BS |
| 003 | 35006 | Premises Signal Transmission | BS | 003 | 37007 | Requirements for Listing | BS |
| 003 | 35008 | Alarm Notification Appliances | BS | 003 | 38001 | Public Fire Alarm Reporting | BS |
| 003 | 36003 | Surveys - Fire Alarm Annunciation | BS | 003 | 38002 | Project Scheduling | BS |
| 003 | 36002 | Low Power Wireless Land Signaling | BS | 003 | 38003 | Contractual Requirements | BS |
| 003 | 37002 | System Features - Hostile Environment | BS | 003 | 38004 | Bid Invitation | BS |

*This column is to be completed by the verifier in accordance with the Statement of Verification in Section 3.

### Section 3 – Statement of Verification

Signature, initials, and date will not be accepted in photocopied form. Please use an ink color other than black.

I certify that I have read the descriptions of the above-listed work elements in the program detail manual, that I have personally initialed work element(s) in Section 2 above, allowing no one to act as my agent, and that my initials attest that I have personally monitored and approved the applicant's repeated and correct completion of the task or application of the knowledge required under a variety of conditions.

Stephen F. Sellers                  Stephen F. Sellers                  BS        12/7/04
Verifier's Name (Please Print)        Verifier's Signature              Verifier's Initials      Date

03/03

2005
NICET TECHNICIAN APPLICATION
PART 3, PERFORMANCE VERIFICATION

Social Security No-Blocked

Name: **Kenneth R. Stephens**  Social Security Number: __

## NICET Technician Application, Part III: Performance Verification
(Please print clearly or type)

**WE** Work Element

### Section 1 –Applicant's Relationship with Verifier

A. Verification will be provided by Mr./Ms. **Stephen F Sellers**
This individual: ☑ has been my immediate supervisor ☑ has never been my immediate supervisor

B. If verification is being provided by a person who has never been your immediate supervisor:
In what way was this person responsible for monitoring and approving your work?

_____

_____

_____

C. During what time period did the verifier oversee your work? from **02 09 92** to **09 30 00**
mo day yr  mo day yr

════════ **The verifier must complete the remainder of this form.** ════════

### Section 2 –Verifier's Personal Information

> Any person who provides work element verifications for this candidate cannot also submit a technician personal recommendation on his or her behalf.

Read carefully the function of the verifier and the work element descriptions in the program detail manual before initialing work elements in the "Selection" table in Part II. Please fill out the information below. Verification will not be accepted unless all items in the "Verifier's Statement" at the bottom of this page are filled in.

Verifier's Name **Stephen F Sellers**   Phone No. **972/470-1379**

Verifier's Job Title **Small Business Install Prom Mgr**

Verifier's Employer **AOT**

Employer Address **2403 Lloyd**
**Carrollton TX 75006**

Summarize your technical experience in the specialty area the applicant is testing: **installed and serviced fire and security systems for over 25 yrs**

List any registrations, certifications or licenses you hold: **RSA**

### Section 3 - Verifier's Statement

I certify that I have been in a position to be responsible for the conduct or results of the applicant's work or that my approval has been required for the products or results of the applicant's work, and that the statement of "Applicant's Relationship with Verifier" at the top of this page is a true representation of my working relationship with the applicant. I further certify that I will verify competent performance or application of knowledge only when I am familiar with the meaning and application of the specific work element, and can, from personal experience, assess the applicant's capability to perform the work element.

Signature **Stephen F Sellers**   Initials **S**   Date **8/27/04**

Signature, initials, and date will not be accepted in photocopied form. Please use an ink color other than black.

**WARNING**
If NICET determines that any verification was obtained from a non-qualified verifier or was given for tasks not actually performed, the institute may permanently deny the certification sought or revoke the certification(s) held by both the applicant and the verifier (if certified). Further, the institute may revoke the verifier's right to submit personal recommendations or verify work element for any NICET certification candidates.

If not sent with the Examination application, please mail this form to NICET, 1420 King Street, Alexandria, VA 22314-2794

2005
NICET TECHNICIAN APPLICATION
PART 3, WORK HISTORY

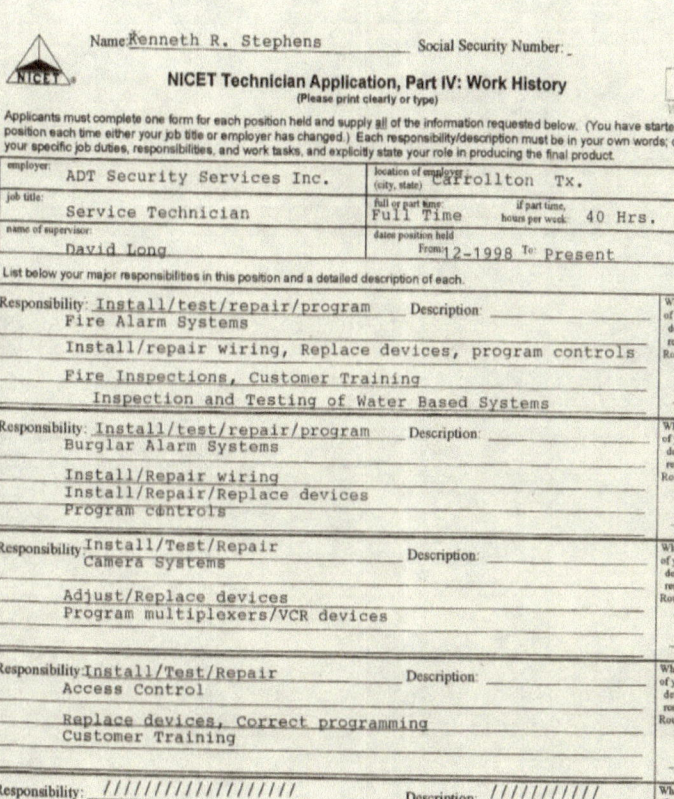

*Kenneth R. Stephens CET*

# TEXAS FIRE ALARM INSPECTOR

Kenneth R Stephens
NICET CERTIFICATIONS

2005-2008

# NICET APPROVAL LETTER

## *APPROVAL LETTER*

Name:                                   Kenneth R Stephens
Date of Award:                     October 28, 2005
Certification Number:          114543
Certification Expire Date:    09/01/2008

It is my pleasure to inform you that you have been awarded certification as follows:

FIRE PROTECTION ENGINEERING TECHNOLOGY/
INSPECTION AND TESTING OF WATER-BASED SYSTEMS/
LEVEL I
FIRE PROTECTION ENGINEERING TECHNOLOGY/FIRE
ALARM SYSTEMS/LEVEL II

If this is your first award of NICET certification, the expiration date shown under your certification number establishes your three-year recertification cycle. If this is an upgraded certification or a certification in a new technical area, your three-year recertification cycle remains the same as previously established. Please refer to NICET Policy No. 30, *Continuing Professional Development*, for rules governing recertification.

Prior to removing the wallet card from this letter, we advise that you make a copy of the letter for your files as the complete letter may be required as proof of certification.

The interest you have shown in your career development by obtaining professional recognition and status through certification is most commendable. On behalf of the Board of Governors, please accept our congratulations and best wishes.

Very truly yours,

Michael A. Clark
General Manager

TEXAS FIRE ALARM INSPECTOR
Kenneth R Stephens
NICET CERTIFICATIONS

2008-2011

# NATIONAL INSTITUTE FOR CERTIFICATION IN ENGINEERING TECHNOLOGIES®

HEREBY CERTIFIES THAT
## Kenneth R Stephens

HAS ATTAINED THE GRADE OF
## LEVEL II

IN FIRE PROTECTION ENGINEERING TECHNOLOGY
FIRE ALARM SYSTEMS

AND RECOGNIZES THAT THROUGH EDUCATION,
EXPERIENCE, AND KNOWLEDGE THIS PERSON HAS
MET THE STANDARDS SET FORTH BY THIS INSTITUTE
Certification Valid through September 1, 2011

CERTIFICATION NUMBER 114543

CHAIRMAN OF THE BOARD OF GOVERNORS, NICET

SPONSORED BY THE NATIONAL SOCIETY OF PROFESSIONAL ENGINEERS

TEXAS FIRE ALARM INSPECTOR

Kenneth R Stephens

NICET CERTIFICATIONS

2011-2014

# NICET APPROVAL LETTER

Name:                                        Kenneth R Stephens

Date of Award:                          September 19, 2011

Certification Number:              114543

Certification Expire Date:       09/01/2014

It is my pleasure to inform you that recertification has been granted as follows:

FIRE PROTECTION ENGINEERING TECHNOLOGY/
INSPECTION AND TESTING OF WATER-BASED SYSTEMS/
LEVEL I
FIRE PROTECTION ENGINEERING TECHNOLOGY/FIRE
ALARM SYSTEMS/LEVEL II

You will find your new wallet card attached to the bottom of this letter. Also enclosed with this letter is your new certificate. Your new three-year period of certification is printed on both your wallet card and your certificate. You will need to accumulate another 90 continuing professional development points to continue your certification beyond this new expiration date.

Prior to removing the wallet card from this letter, we advise that you make a copy of the letter for your files as the complete letter may be required as proof of certification.

The interest you have shown in your career development by obtaining professional recognition and status through certification is most commendable. On behalf of the Board of Governors, please accept our congratulations and best wishes.

Very truly yours,


Michael A. Clark
Chief Operating Executive

*Kenneth R. Stephens CET*

# KENNETH R STEPHENS
# NICET CERTIFICATION-

# LEVEL 1
# WATER BASED INSPECTIONS

## NATIONAL INSTITUTE FOR CERTIFICATION IN ENGINEERING TECHNOLOGIES®

Providing Certification Programs Since 1961

BE IT KNOWN THAT

**Kenneth R Stephens**

IS HEREBY AWARDED CERTIFICATION AT

LEVEL I

IN FIRE PROTECTION ENGINEERING TECHNOLOGY
INSPECTION AND TESTING OF WATER-BASED SYSTEMS

BASED UPON SUCCESSFUL DEMONSTRATION OF REQUISITE KNOWLEDGE,
EXPERIENCE AND WORK PERFORMANCE AS SET FORTH BY THIS INSTITUTE.

Certification Valid through September 1, 2014

CERTIFICATION NUMBER 114543

CHAIRMAN OF THE NICET BOARD OF GOVERNORS
A DIVISION OF THE NATIONAL SOCIETY OF PROFESSIONAL ENGINEERS

**50**th
Anniversary
1961—2011

*Kenneth R. Stephens CET*

Kenneth R Stephens

NICET LEVEL 2

FIRE ALARM SYSTEMS

**NATIONAL INSTITUTE FOR CERTIFICATION IN ENGINEERING TECHNOLOGIES®**

*Providing Certification Programs Since 1961*

BE IT KNOWN THAT

**Kenneth R Stephens**

IS HEREBY AWARDED CERTIFICATION AT

LEVEL II

IN FIRE PROTECTION ENGINEERING TECHNOLOGY
FIRE ALARM SYSTEMS

BASED UPON SUCCESSFUL DEMONSTRATION OF REQUISITE KNOWLEDGE, EXPERIENCE AND WORK PERFORMANCE AS SET FORTH BY THIS INSTITUTE.

Certification Valid through September 1, 2014

CERTIFICATION NUMBER 114543

CHAIRMAN OF THE NICET BOARD OF GOVERNORS
A DIVISION OF THE NATIONAL SOCIETY OF PROFESSIONAL ENGINEERS

**50**th
*Anniversary*
*1961—2011*

*Kenneth R. Stephens CET*

# TEXAS FIRE ALARM INSPECTOR

Kenneth R Stephens

# PREVIOUS FIRE ALARM LICENSES

# STATE OF TEXAS

# TEXAS FIRE ALARM LICENSE

TEXAS COMMISSION ON FIRE PROTECTION
STATE FIRE MARSHAL'S OFFICE
FIRE ALARM LICENSE RENEWAL APPLICATION

| C of R Number | License Number | Expiration Date | Renewal Fee |
|---|---|---|---|
| ACR-1036 | FAL-2688 | 2/27/96 | $200.00* |

Please make any necessary corrections on the following addresses.

Home address of licensee:

STEPHENS, KENNETH R
25 HIGHLAND PK.VL.100-108
DALLAS, TX 75205

Mailing address of employer:

ALERT CENTRE, INC.
5800 SOUTH QUEBEC STREET

ENGLEWOOD, CO 80111

* If you are renewing a license with more than one employer, the fee for each additional license is $20.00 after the first $200.00 is paid.

Please note that LATE FEES will be required for renewal applications that are not postmarked by the expiration date and also for those that are not deemed complete, according to both statute and rules, before the expiration of the license or within the 30-day notice period allowed by the fire alarm statute.

Have you been convicted of a felony or a misdemeanor within the last two years?

( ) YES        (X) NO

If "YES", give details for all convictions within the last two years. Your name is being submitted to the Department of Public Safety for criminal history review. Failure to report all convictions could result in denial of your license.

Original signature of renewal applicant: *Kenneth R. Stephens*

STEPHENS, KENNETH R

Our firm has investigated this applicant's character and reputation for honesty and trustworthiness and we are satisfied that the applicant will act in good faith toward the public as an employee or agent of this firm. No fact or condition which would disqualify the applicant from receiving a license is known to us.

Signature of authorized representative of firm:

Original signature _____        Date  01-10-1996

Printed name  JEFFREY B. WINTER_____        Title  COMPLIANCE ADMINISTRATOR

Any fraudulent representation on this application may be cause for denial, suspension, or revocation of a license.

Check or money order for the renewal fee should be made payable to the Texas Commission on Fire Protection and mailed with this complete application to the:

Texas Commission on Fire Protection
Licensing Administration
P. O. Box 2286
Austin, Texas 78768-2286

SLL 028
(Rev. 11/14/95)

12/14/95

# THE TEXAS FIRE ALARM INSPECTOR

Kenneth R Stephens

## PREVIOUS TEXAS FIRE ALARM LICENSES

ISSUED TO:

STEPHENS, KENNETH R
25 HIGHLAND PK.VL.100-108
DALLAS, TX 75205

LICENSE NUMBER:  FAL-2688

EXPIRATION DATE:  2/27/98
REGISTRATION NO:  ACR-1036

TEXAS COMMISSION ON FIRE PROTECTION

FIRE ALARM TECHNICIAN LICENSE

GARY L. WARREN, SR., EXECUTIVE DIRECTOR
G. MIKE DAVIS, STATE FIRE MARSHAL

MAIL TO:

✓ ALERT CENTRE, INC.
5800 SOUTH QUEBEC STREET
ENGLEWOOD, CO 80111

✓ 1/19/96

ISSUED BY:

*Mary Lou Ramirez*

MARY LOU RAMIREZ, PROGRAM DIRECTOR
LICENSING ADMINISTRATION

---

ISSUED TO:

STEPHENS, KENNETH R
25 HIGHLAND PK.VL.100-108
DALLAS, TX 75205

LICENSE NUMBER:  FAL-2688

EXPIRATION DATE:  2/27/98
REGISTRATION NO:  ACR-1030

TEXAS COMMISSION ON FIRE PROTECTION

FIRE ALARM TECHNICIAN LICENSE

GARY L. WARREN, SR., EXECUTIVE DIRECTOR
G. MIKE DAVIS, STATE FIRE MARSHAL

MAIL TO:

✓ HAWK SECURITY SERVICES
5718 AIRPORT FREEWAY
FORT WORTH, TX 76117

✓ 10/28/96

ISSUED BY:

*Mary Lou Ramirez*

MARY LOU RAMIREZ, PROGRAM MANAGER
LICENSING ADMINISTRATION

---

ISSUED TO:

STEPHENS, KENNETH R
25 HIGHLAND PK.VL.100-108
DALLAS, TX 75205

LICENSE NUMBER:  FAL-2688

EXPIRATION DATE:  2/27/2000 ✓
REGISTRATION NO:  ACR-1328

TEXAS DEPARTMENT OF INSURANCE
STATE FIRE MARSHAL'S OFFICE
FIRE ALARM TECHNICIAN LICENSE

MAIL TO:

✓ SMITH ALARM SYSTEMS, INC.
ATTN: CHRIS DIERKS
7777 CARPENTER FREEWAY

G. MIKE DAVIS, STATE FIRE MARSHAL

ISSUED BY:

*Mary Lou Ramirez* ct

82

Kenneth R Stephens

PREVIOUS TEXAS FIRE ALARM LICENSES.

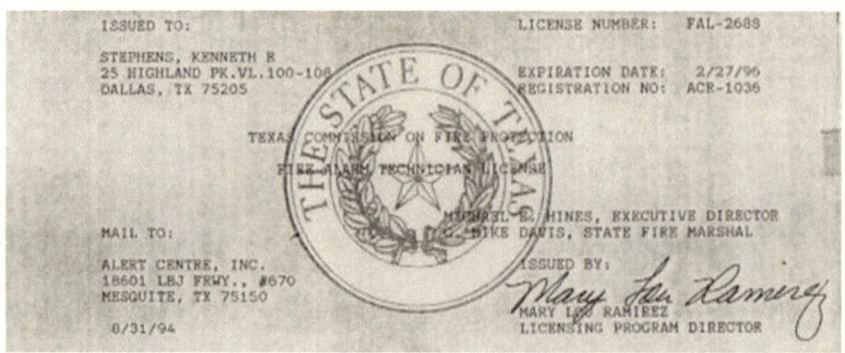

84

Kenneth R Stephens

Previous Texas Fire Alarm Licenses.

Issued To:
**STEPHENS, KENNETH R**
655 E ROYAL LN
IRVING TX 75034

License Number
**FAL-2688**
Expiration Date: 02-27-2012

STATE OF TEXAS

**FIRE ALARM TECHNICIAN LICENSE**
*TEXAS DEPARTMENT OF INSURANCE*
*STATE FIRE MARSHAL'S OFFICE*

DATE ISSUED: November 20, 2010

PAUL MALDONADO, STATE FIRE MARSHAL

SFO81 Rev.120

STATE OF TEXAS

ALARM RECORDS

*Kenneth R. Stephens CET*

# TEXAS FIRE ALARM INSPECTOR

Kenneth R Stephens

## ALARM COMPANY WORK HISTORY

## STATE OF TEXAS

TEXAS PRIVATE SECURITY BUREAU

ALARM INSTALLER LICENSE (BURGLAR ALARM)

Person Details

## Person Details

Status:         Approved

Name:           STEPHENS, KENNETH

Gender:         M

DOB:

## License Details

| Alarm Installer | 12/10/2002 | 12/10/2006 | Active |
| --- | --- | --- | --- |

## Employment Details

| A-AMERICAN ALARM CO B03977 | B03977 | Alarm Installer | |  |
| --- | --- | --- | --- | --- |
| ADT SECURITY SERVICES, INC. | B00536 | Alarm Installer | 12/9/1998 | |
| ADT SECURITY SERVICES, INC. | B00536 | Alarm Installer | | |
| ADT SECURITY SERVICES, INC. | B00536 | Alarm Installer | | |
| ADVANCED ENTRY SYSTEMS, INC. B04946 | B04948 | Alarm Installer | | |
| ALERT CENTRE, A DIVISION OF ADT SECURITY B05703 | B05703 | Alarm Installer | | |
| PROGRESSIVE CONCEPTS, INC DBA HAWK SECURITY SERVICES | B04632 | Alarm Installer | | |
| SMITH ALARM SYSTEMS | B00380 | Alarm Installer | | |
| ??? of Dallas Inc. | B00542 | Alarm Installer | | |

## Training Details

| | |
|---|---|
| | |

Note: DOB is blocked out.

*Kenneth R. Stephens CET*

Kenneth R Stephens

ADT

5 year and 10 year Awards

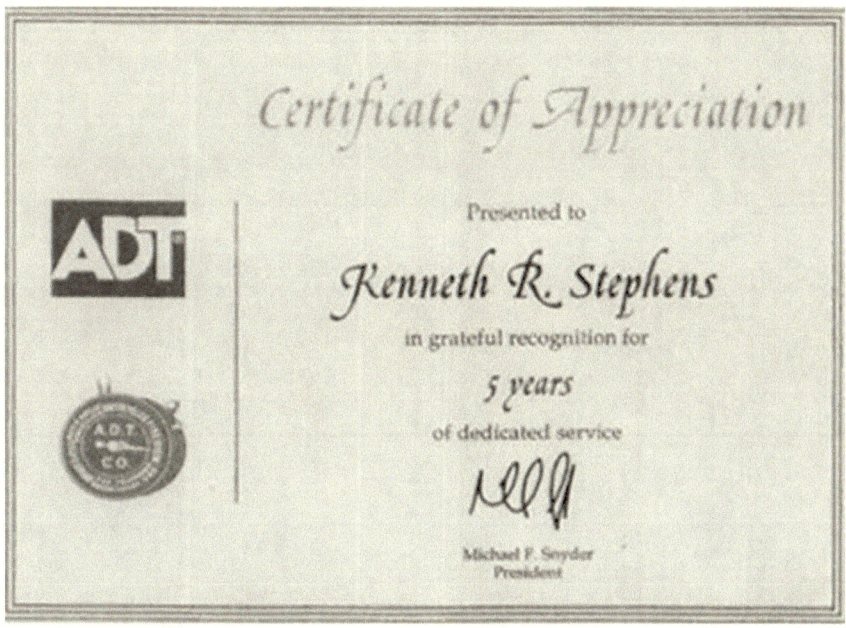

*Kenneth R. Stephens CET*

# NATIONAL BURGLAR and FIRE ALARM ASSOCIATION
# NATIONAL TRAINING SCHOOL

TEXAS FIRE ALARM INSPECTOR


NATIONAL BURGLAR & FIRE ALARM ASSOCIATION
NATIONAL TRAINING SCHOOL

CERTIFIED ALARM TECHNICIAN LEVEL I

SMITH ALARM SYSTEMS

*Kenneth R. Stephens CET*

Kenneth R Stephens
Certified Alarm Technician
NBFAA

National Training School
This is to certify that

## Kenneth R Stephens

has successfully completed the Level I training course and is named

### CERTIFIED ALARM TECHNICIAN

of the

## *National Burglar and Fire Alarm Association*

Date    **September 22, 1991**

Director    President

Kenneth R Stephens
SMITH ALARM SYSTEMS
APPRENTICE
INSTALLER/ REPAIRMAN
LEAD REPAIRMAN

KENNETH R STEPHENS
SMITH ALARM SYSTEMS
1980-1992
1997-1998

*Kenneth R. Stephens CET*

Kenneth R Stephens

MANUFACTURERS CERTIFIED TRAINING

FIRELITE ALARMS

*Kenneth R. Stephens CET*

Kenneth R Stephens

WATER BASED INSPECTIONS TRAINING

**Construction Education Foundation, Inc.**

*Hereby certifies that*

## Kenneth R. Stephens

*Has successfully completed 8 hours of training in*

## Fire Protection Inspection Evaluation & Corrective Action II

*Presented by*
*CEF and North Lake College*

*Issued on this 22nd day of September, 2010*

*Ken Polk*
CEF Chairman of the Board

*Kenneth R. Stephens CET*

Kenneth R Stephens

# COMPUTER TRAINING (ADT MASTERMIND PROGRAM)

## CERTIFICATE OF COMPLETION

This certifies that

**Ken Stephens**

has successfully completed the course

**Beginning Windows 95**

This is an 8-hour course

*Ann Bailey*

February 20, 1999
CPE Sponsor ID#08069
CEU 0.7

New Horizons
Computer Learning Centers

CERTIFIED TRAINING

"MASTERMIND" COMPUTER PROGRAM

*Kenneth R. Stephens CET*

# CERTIFIED TRAINING
# "MASTERMIND" COMPUTER PROGRAM

Kenneth R Stephens

U.S.     MARSHALS SERVICE CLEARANCE
         Public Trust
U.S.     DEPARTMENT of DEFENSE SECURITY CLEARANCE
         DoD

# TEXAS FIRE ALARM INSPECTOR
Kenneth R Stephens

## SECURITY CLEARANCES

ADT FEDERAL SYSTEMS DIVISION

Stephens, Kenneth R

| From: | USMS Clearance          Sent: Tue 5/6/2008 12:10 PM |
|---|---|
| To: | Stephens, Kenneth R |
| Cc: | Long, David S (Carrollton) |
| Subject: | Security Clearance Approval Notice of Kenneth R Stephens |
| Attachments: | Kenneth Ray Stephens.pdf (98KB) |

The investigation report on the above subject has been determined that the applicant is suitable for employment.

Therefore, the U.S. Marshals Service Headquarters has APPROVED Kenneth R Stephens for access to U.S. Marshals Service facilities to perform contracted work.

Town #: 0023

Any questions should be directed to the email address usmsclearance@adt.com

Congratulations,

ADT/USMS Project Office

—

Craig West
ADT Security Services, Inc.
Federal Systems Division
3601 Eisenhower Avenue, 3rd Floor
Alexandria, VA. 22304
Direct Office Line: 703-317-4215

# THE TEXAS FIRE ALARM INSPECTOR
# WORKSHEETS

*Kenneth R. Stephens CET*

# THE TEXAS FIRE ALARM INSPECTOR

From: Bryant, Vanessa
Sent: Thursday, December 18, 2003 12:52 PM
To: Sanders, D'Andrea Bernard
Subject: Ken Stephens

Ken Stephens asked me to email his DOD security clearance status information, please see below.

Full Name: Kenneth R  Stephens
Clearance Date: 2/6/2003
Clearance Level: Secret

Please let me know if you require additional information.

Sincerely,

Vanessa Bryant
ADT Security
Asst. FSO & Licensing Administrator

Issued March 19, 1999                                    Revised November 2, 2000

# Notice

### EMPLOYEE – STATUS CHANGE OR ADVERSE INFORMATION NOTICE AND REPORT

This employee, _Stephens, Kenneth_ holds a security clearance with the Department of Defense. The National Industrial Security Procedure Operating Manual (NISPOM) Section 3 requires that contractors (ADT) report any changes which may affect personnel security clearances to the Cognizant Security Agency (CSA). Any change in this employee's status which may affect his/her clearances should be reported to the Local ADT Facility Security Officer, who will then make a report to the appropriate CSA office.

# EASTFIELD COLLEGE

### AWARDS

1.8

#### CONTINUING EDUCATION UNIT(S)

TO

KENNETH R. STEPHENS

FOR SATISFACTORY COMPLETION OF THE REQUIREMENTS OF THE

## CONTINUING EDUCATION COURSE

TELEPHONE INSTALLATION & REPAIR

*AWARDED AT DALLAS, TEXAS, THE* 25th *DAY OF* MARCH 19 85

*Eleanor Ott*

**President**

**Instructor**

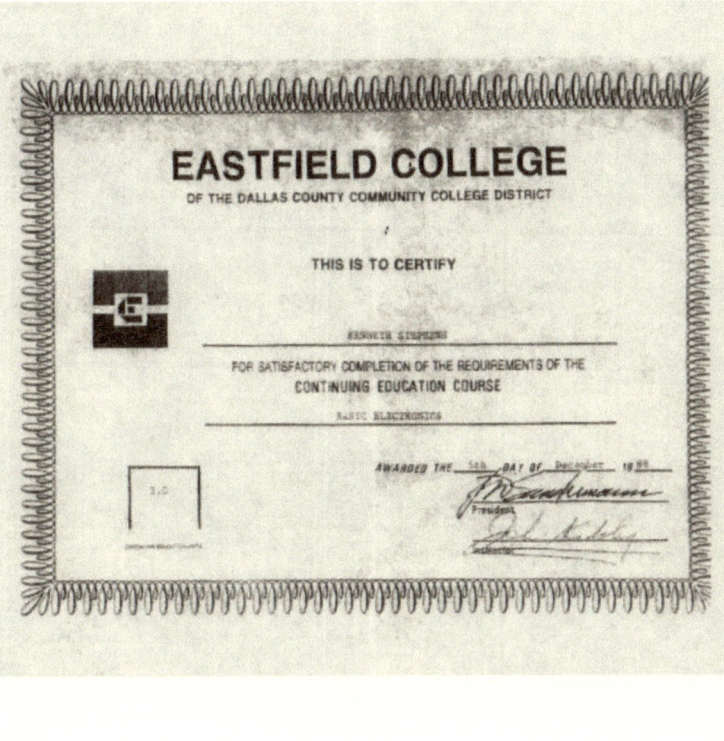

# EASTFIELD COLLEGE

### OF THE DALLAS COUNTY COMMUNITY COLLEGE DISTRICT

**THIS IS TO CERTIFY**

KENNETH STEPHENS

FOR SATISFACTORY COMPLETION OF THE REQUIREMENTS OF THE
**CONTINUING EDUCATION COURSE**

BASIC ELECTRONICS

AWARDED THE 5th DAY OF DECEMBER 19 85

ESSENTIAL TRAINING
            BUSINESS COMMUNICATIONS

A B C

ACCURACY

BREVITY

CLARITY                                    .


        ACCURACY

        CORRECT INFORMATION.
        EVERYTHING YOU SAY IS :
            a. ACTUALLY TRUE
            b. CONFIRMED REPORT
            c. NO OMMISSIONS ( YOU OMMITTED NOTHING).
            d. REPORTED FAIRLY ( NO BIAS).
            e. AVOID BROAD SWEEPING STATEMENTS
               THESE CAN BE EASILY CHALLENGED.

         BREVITY
         BE BRIEF
            a. USAGE
                1. SIMPLE
                2. DIRECT
                3. CAREFULLY CHOSEN WORDS (sentences or paragraphs)
            b. GET TO THE POINT

         CLARITY
         BE CLEAR
            a. PUT YOUR THOUGHTS IN LOGICAL ORDER.
            b. CHOOSE WORDS THAT GET THE MEANING ACROSS;
                1. QUICKLY
                2. EFFICIENTLY
            c. IMPROVE VOCABULARY
            d. CONVERSATIONAL
            e. REREAD;
               SEE IF YOU UNDERSTAND WHAT YOU HAVE WRITTEN.


REFERENCES: MULTIPLE COLLEGE TEXTBOOKS

# About the Author

Kenneth R Stephens is a NICET certified, State of Texas licensed Fire Alarm Inspector based in Dallas County, Texas. Kenneth has worked for his current company (an international corporation) for fifteen years. Kenneth worked for his previous company for thirteen years and it was acquired by his current company. Kenneth has worked in the alarm industry in Dallas Texas since 1980 and every company he has worked for was acquired by his current company. Kenneth is NICET certified in Fire Protection Engineering Technology: Inspection and Testing of Water Based Systems (NFPA 25) Level 1 and Fire Protection Engineering Technology: Fire Alarm Systems (NFPA 72) Level 2. Kenneth holds a State of Texas Fire Alarm License (FAL-Commercial Technician) which he has held at this level since 1991. Kenneth was awarded Certified Alarm Technician Level I by the National Burglar and Fire Alarm Association; National Training School in 1991. Kenneth has been a registered Alarm Installer with the State of Texas (currently under the Texas Dept. of Public Safety-Private Security Bureau) since 1980. Kenneth attended the Eastfield Community College Technical Certification Program (DCCCD). His College certificates are: Telephone Installation and Repair-1985. Basic Electronics-1988. Introduction to Locksmithing- 1990. Kenneth entered the alarm industry in Dallas Texas in 1980 and served an alarm apprenticeship of one year working for his company in cooperation with his union, the International Brotherhood of Electrical Workers.